Curious Creatures

AYE-AYES

GAIL TERP

BLACK RABBIT BOOKS

Bolt is published by Black Rabbit Books
P.O. Box 227, Mankato, Minnesota, 56002
www.blackrabbitbooks.com

Marysa Storm, editor; Michael Sellner, designer and photo researcher

Library of Congress Cataloging-in-Publication Data
Names: Terp, Gail, 1951- author.
Title: Aye-ayes / by Gail Terp.
Description: Mankato, Minnesota : Black Rabbit Books, [2023] | Series: Bolt. Curious creatures | Includes bibliographical references and index. | Audience: Ages 8-12 | Audience: Grades 4-6 | Summary: "Aye-ayes are animals unlike any other, and curious and reluctant readers alike will be captivated by these exceptional creatures' life cycle, habitats, diet, and threats to survival through carefully leveled text, detailed infographics that emphasize visual literacy, and engaging, colorful photography"– Provided by publisher.
Identifiers: LCCN 2020016550 (print) | LCCN 2020016551 (ebook) | ISBN 9781623105648 (hardcover) | ISBN 9781644664896 (paperback) | ISBN 9781623105709 (ebook)
Subjects: LCSH: Aye-aye–Juvenile literature.
Classification: LCC QL737.P935 T47 2023 (print) | LCC QL737.P935 (ebook) | DDC 599.8/3-dc23
LC record available at https://lccn.loc.gov/2020016550
LC ebook record available at https://lccn.loc.gov/2020016551

Image Credits

Alamy: Alison Wright, 11; Chris Hellier, 3, 22–23, 31; Chris Mattison, 25; Eric Nathan, 25; imageBROKER, 16–17, 25, 32; Nature Picture Library, 25; Neil Bowman, 12–13; Rob Cousins, 18–19; Suzanne Long, 25; Blue Planet Archive: Thorsten Negro / imageBROKER, 26; Minden Pictures: Chien Lee, 6–7, 8–9; Mark Carwardine, 1, 28–29; Nick Garbutt, 4–5; Suzi Eszterhas, 6–7; Nat Geo Creative: Frans Lanting, 21; Newscom: Denver Zoo / MEGA, 19; Nick Garbutt, 11; Science Source: Thomas & Pat Leeson, 22; Shutterstock: Amyphotostory, 20, 27; Dennis van de Water, 26; Eric Isselee, 25; Hugh Lansdown, 12–13; KatBranchArt, 14–15; Pierre-Yves Babelon, 12–13; Pisut chounyoo, 25; The Duke Lemur Center (http://lemur.duke.edu): David Haring, Cover, 11, 21

Contents

CHAPTER 1

An Aye-Aye in Action

It is nighttime. An aye-aye creeps along a tree branch. As it moves, it taps the branch with its extra-long middle finger. The aye-aye listens. When its huge ears hear movement inside the branch, it stops crawling. It chews at the spot to make a hole. Wood chips fly, but this is no problem for the **primate**. The aye-aye's third eyelid protects its eyes.

How Big Is an Aye-Aye?

TAIL LENGTH
17 TO 24 INCHES
(43 to 61 centimeters)

Finding Food

The aye-aye stops chewing. It's now ready to nab its food. It sticks its long finger into the hole. It pokes around. There! Its finger senses an insect **larva**. With a practiced movement, the aye-aye scoops out the grub. It pops the larva in its mouth and moves on to find more.

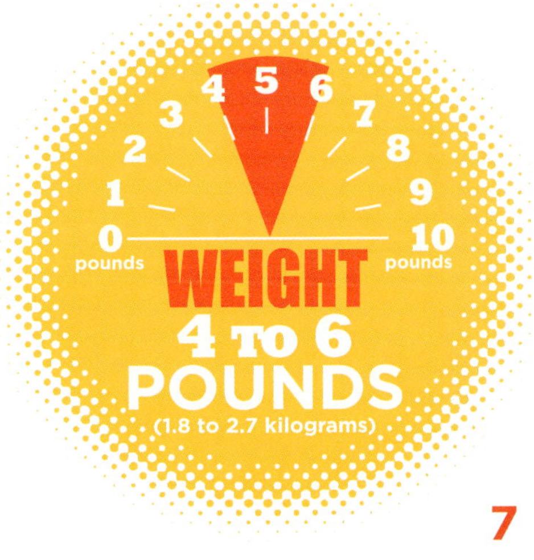

PARTS OF AN AYE-AYE

YELLOW-ORANGE EYES
CLAWS
BUSHY TAIL

CHAPTER 2

FOOD and Home

Aye-ayes eat more than just insect larvae. These primates also eat seeds, fruits, and nuts. They use their teeth to open the nuts' shells. Then they reach in with their middle fingers. With them, they pull out the tasty insides. They also use their fingers to scoop out the insides of coconuts and other fruit.

Aye-ayes are the only primates with teeth that never stop growing.

scooping

digging

eating

Aye-Aye Habitats

RAIN FORESTS

SCRUB FORESTS

FARMS

At Home in the World

Aye-ayes only live in Madagascar. Most make their homes in the island's rain forests. But some live in trees near farms. They eat the yummy **crops**.

Aye-ayes are nocturnal. They spend the day sleeping in nests of twigs and leaves. At night, they search the trees for food.

WHERE AYE-AYES LIVE

Aye-Aye Range Map

MADAGASCAR

CHAPTER 3

The Family

Aye-ayes mostly live alone. But sometimes, they form pairs or small groups of up to four. These groups travel and eat together. Aye-ayes are most **social** when they **mate**. Mating can happen any time of the year.

Aye-ayes make sounds to **communicate**. Some sounds mean danger is near. Others are used when aye-ayes fight. Mothers and their young use sounds to help them find each other.

Newborns have green eyes. Their eyes change color as they grow up.

A Baby is Born

Females give birth to one baby at a time. They wait two to three years before they have their next one. Newborns stay in their mothers' nests for about two months. There, they stay safe and nurse. Then they start to explore. But they aren't ready to be alone. They still need their mothers to take care of them.

Growing Up

Outside the nest, young aye-ayes have much to learn. At first, they play. They learn how to move about the trees and use their long fingers. When they're ready, their mothers will show them how to get food. By age two, aye-ayes are ready to live on their own.

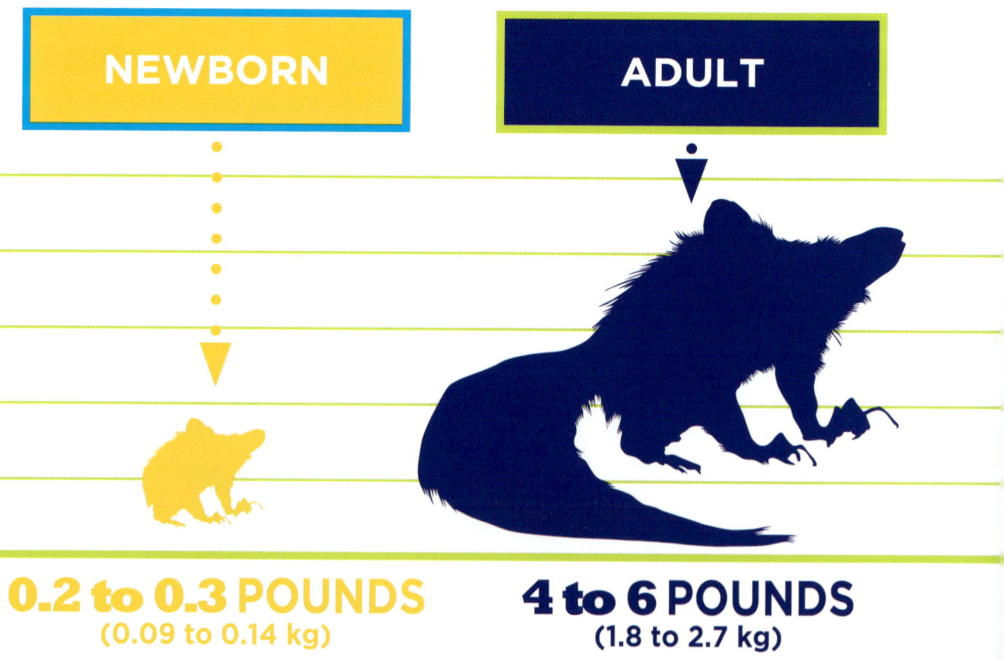

COMPARING WEIGHTS

By the Numbers

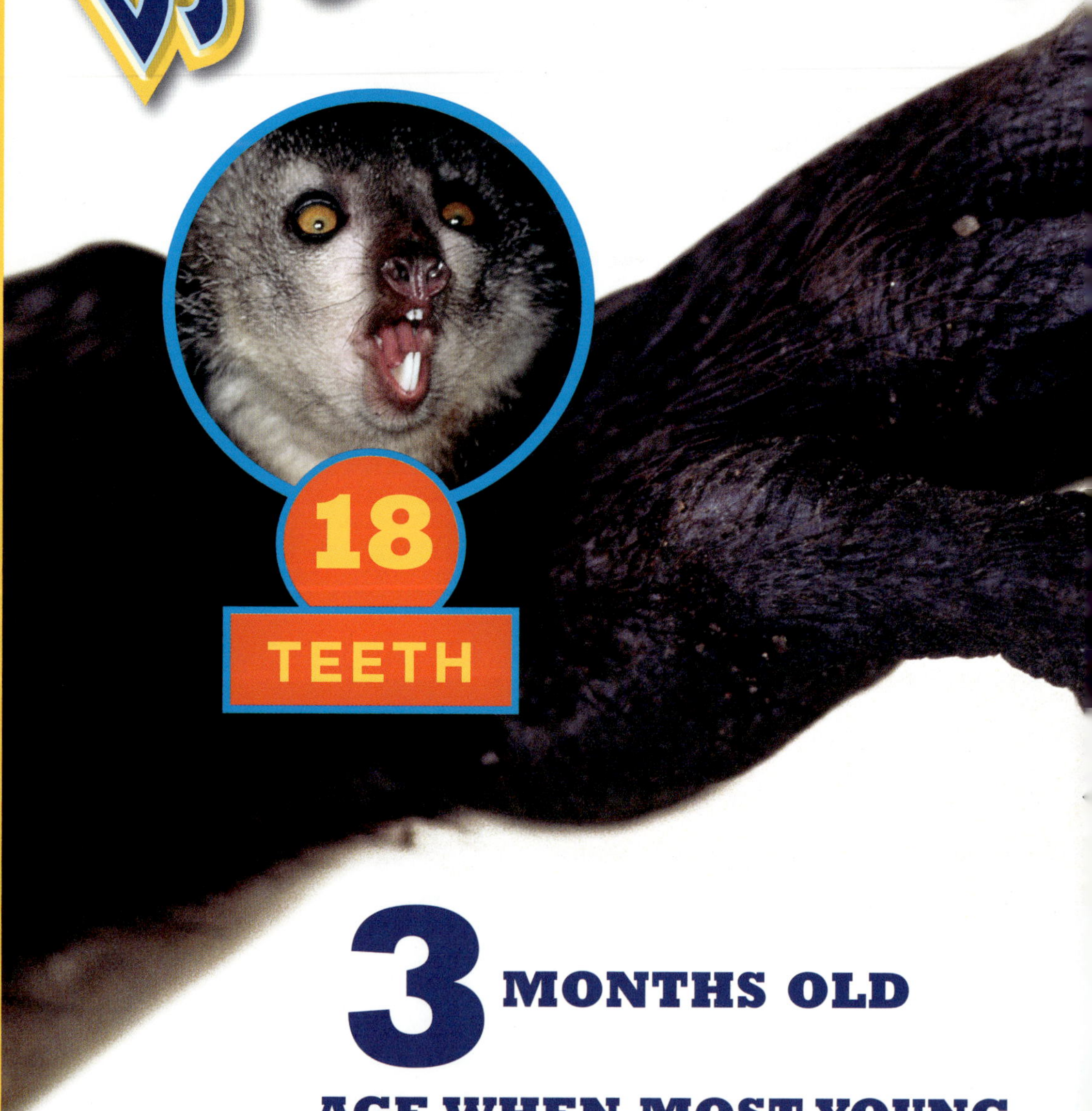

18
TEETH

3 **MONTHS OLD**

AGE WHEN MOST YOUNG AYE-AYES FIRST EAT SOLID FOOD

up to 24 YEARS
LIFE SPAN IN ZOOS

about 10 times per second

HOW FAST THEY CAN TAP THEIR LONG FINGER

CHAPTER 4

Danger!

Fossas are the main **predator** of adult aye-ayes. Like aye-ayes, these catlike animals only live in Madagascar. They can grow up to 6 feet (1.8 meters) long. Other predators, such as snakes and hawks, hunt young and smaller aye-ayes.

Aye-Aye Food Chain

This food chain shows what aye-ayes eat.
It also shows what eats aye-ayes.

Human Threat

Humans are the biggest danger to aye-ayes. They hunt aye-ayes for food. Some people also kill aye-ayes because they believe the animals are bad luck.

Aye-ayes' largest threat comes from habitat destruction. Humans cut down forests. Aye-ayes have fewer places to live. They have fewer trees to get food from.

Aye-ayes are on the list of the world's 25 most endangered primates. No one knows how many there are. This uncertainty is because people rarely see them.

Protecting Aye-Ayes

Researchers study these endangered animals. What they learn about aye-aye behavior and habitat will help people protect them.

Aye-ayes play a big part in their **ecosystems**. They plant new trees by dropping seeds. They also eat insects that damage trees. These unique animals must be protected.

communicate (kuh-MYU-nuh-kayt)—to share information, thoughts, or feelings so they are understood

crop (KROP)—a plant that is grown by farmers

ecosystem (E-co-sys-tum)—a community of things living in one place

endangered (in-DAYN-jurd)—close to becoming extinct

larva (LAR-vuh)—the wormlike form of an animal that hatches from an egg

mate (MAYT)—to join together to produce young

predator (PRED-uh-tuhr)—an animal that eats other animals

primate (PRI-mayt)—any member of the group of animals that includes humans, apes, and monkeys

social (SO-shul)—liking to be with and communicate with others

LEARN MORE

BOOKS

Bassier, Emma. *Aye-Ayes.* Weird and Wonderful Animals. North Mankato, MN: DiscoverRoo, an imprint of Pop!, 2020.

Levy, Janey. *The Unsightly Aye-Aye.* Nature's Freak Show: Ugly Beasts. New York: Gareth Stevens Publishing, 2020.

Shumaker, Debra Kempf. *Peculiar Primates: Fun Facts about These Curious Creatures.* Philadelphia: Running Press Kids, 2022.

WEBSITES

Aye-Aye
kids.nationalgeographic.com/animals/mammals/aye-aye/

Aye Aye Animal Facts
a-z-animals.com/animals/aye-aye/

Aye Aye Lemur
www.dkfindout.com/us/animals-and-nature/primates/aye-aye/

INDEX